SELF-SUFFICIENCY

FOR A SUSTAINABLE AUSTRALIAN FUTURE

Self-Sufficiency for a Sustainable Australian Future examines the notion that reliance on economic materialism for growth and consumer wellbeing is not only a flawed concept but also directly leads to environmental catastrophe and climate change. Self-sufficiency—do-it-yourself homemaking—offers real solutions to these problems.

By growing, cooking and preserving, building, raising animals, producing energy, and making and doing, we can actively practice an authentic lifestyle: a lifestyle based on creativity, self-help, independence, self-expression, and freedom—elements that are, and perhaps always will be, lacking or impossible in the mainstream.

Amidst the complexity that is modern life, this book offers a message of hope. No matter your material circumstances, your geographical location or your age, everyone can be a little bit self-sufficient. The more self-sufficient you are the better you will feel and freer you will be.

Why not give it a go?

Amanda McLeod received her doctorate in consumer history from Monash University in 2004. Her first book, *Abundance: Mass Consumption in Postwar Australia* was published in 2007. She has more than forty publications that have appeared in a diverse range of sources including *Journal of Historical Research in Marketing, History Australia, Australian Policy and History* and *Business History*. Her second book *Simple Living in History: Pioneers of the Deep Future* was published to critical acclaim in 2014. She publishes regularly at historicalperspectives.com.au.

SELF-SUFFICIENCY

FOR A SUSTAINABLE AUSTRALIAN FUTURE

AMANDA MCLEOD

PAMPHLETEER

Pamphleteer is an Australian Scholarly imprint.

First published 2017 by Australian Scholarly Publishing Pty Ltd
7 Lt Lothian St Nth, North Melbourne, Vic 3051
TEL: 03 9329 6963 FAX: 03 9329 5452
EMAIL: aspic@ozemail.com.au WEB: scholarly.info

ISBN 978-1-925588-59-0

*A society in which consumption has to be artificially stimulated in order to keep production going is a society founded on trash and waste, and such a society is a house built upon sand.**

* Dorothy L. Sayers, 'Why Work?' in *Creed or Chaos and Other Essays in Popular Theology*, London: Methuen, 1947, p. 47; also quoted in Packard, *The Waste Makers*, London: Longmans, 1960.

CONTENTS

ACKNOWLEDGEMENTS

A number of people challenged and supported the ideas that eventually resulted in this little book. I am indebted to: Linda Barclay, Toby Handfield, Michael Allaby, Samuel Alexander, Susan Brinksma, Kathy Lothian and Mark Dibben. Thanks must also be given to my publisher, Nick Walker and the team at Australian Scholarly Publishing who saw the value in this work. And, finally to my boy Owen McLeod-Agland, always ready to impart wisdom and share his good humour. This book is for him.

INTRODUCTION

We have reached the 'age of consequences'.*

We are the bringers of devastation and destruction.

As the result of more than two hundred years of consumer capitalism and the burning of fossil fuels, carbon emissions from mass transportation, emissions from intensive animal production and agriculture, deforestation and the use of chemical fertilizers, our climate is changing exponentially. The consequences are dire and many.

Anthropogenic (human-induced) climate change causes global warming, sea level rise, ocean acidification, desertification, the melting of glaciers, droughts and crop failures, diseases, more frequent hurricanes and wildfires, heat waves, severe precipitation, longer or shorter seasons, mass animal extinction, economic collapse, air related illnesses, decreased population, the disappearance of entire countries, mass failure of energy systems, lack of fresh water, the

* Jared P. Scott (director), *The Age of Consequences*, documentary, 2016.

disappearance of coral reefs, and mass human extinction. It also brings war.

Those of us in developed countries are to blame. We must take responsibility for our part in the problem.

We must do something about the climate emergency, now.

We must lower our carbon footprint, reduce food miles and use new and more efficient technology that does not harm the environment and cause further climate change.

But we need to do more than this.

We also need to significantly reduce our reliance on consumerism to deliver abundance and prosperity. We need to be more independent and self-sufficient.

***

Self-sufficiency stems from the desire to produce as much of one's needs as possible. By growing, cooking and preserving, building, raising animals, producing energy, and making and doing, we can actively practice an authentic lifestyle: a lifestyle based on creativity, self-help, independence, self-expression, and freedom – elements that are, and perhaps always will be, lacking or impossible in the mainstream. Self-sufficiency's advantages for combating climate change and other negative effects of capitalism are the subjects of this essay.

Ernst Friedrich Schumacher, the author of the ground-breaking *Small is Beautiful: A Study of Economics as if People Mattered*, published in 1973, wrote that the economic system based on capitalism was corrupt and needed to be changed. The problem for him stemmed from the 'problem of production' and man's attitude to nature. Such a problem, Schumacher explained, was one that could not be ignored because it was based and entirely dependent on the burning of fossil fuels. Put simply, fossil fuels were a finite resource that could not be replaced.*

More than forty years later not much has changed. Despite the negative consequences, not least the devastating impact on the natural environment, much of our economic system is still based on the burning of fossil fuels. Yet, there are many reasons to be concerned about fossil fuels beyond their finite nature. Fossil fuels – on which the capitalist system heavily relies – are directly responsible for climate change.

Alternatives need to be found to replace such a system. A system that is all but broken. In addition to the negative environmental aspects of consumer capitalism, there have always been those who thought there was a better kind of life than one based on consumer choices. Some chose alternative lifestyles of various sorts: from city and country intentional communities to artist or spiritually focused

* E.F. Schumacher, *Small is Beautiful: A Study of Economics as if People Mattered*, 1973, London: Blond and Briggs, pp. 10–11.

communes. Others saw the problem as one of unbridled growth and looked for alternatives to the mainstream by taking up various forms of protest. And an increasingly dissatisfied minority looked to self-sufficiency as a panacea to society's woes.

It is perhaps no surprise then, that Schumacher wrote the foreword to John Seymour's *Complete Guide to Self-Sufficiency* in 1976. By this time, the long economic boom of the postwar period had come to an end. But the reliance on economic growth for prosperity remained. In linking his critique of economics to self-sufficiency, Schumacher argued that never before had people been so dependent on the 'system' and less able to do things for themselves. There were serious consequences for being too dependent: 'What if there is a hold-up, a breakdown, a strike, or unemployment?' He accepted that the State could no longer be relied upon to provide a safety net; even if it existed, people could fall through the cracks. 'Why can't they help themselves?', Schumacher asked. For him the answer was all too obvious: 'they would not know how to; they have never done it before and would not even know where to begin'.* During the affluent years people had simply lost the skills to be independent and resourceful. Some of the answers, of course, lay in Seymour's how-to manual and self-sufficiency. The independence and freedom that flow

---

* John Seymour, *The Complete Book of Self-Sufficiency*, London: Faber and Faber, 1976, p. 6.

from self-sufficiency is my concern here.

Capitalism was thought to be the solution to poverty and the bringer of prosperity. Materialism, capitalism's natural bedfellow, promised to maintain full employment, to build and furnish houses and to elevate the majority of people to the middle class. And it was to a certain extent successful. Capitalism and materialism, of course, have had their strident critics. Despite this, the majority of pundits have lauded their importance for increasing growth and consumer wellbeing. Some were more enthusiastic than most.

When Joe Hockey told the Institute of Economic Affairs in London in 2012 that the 'age of entitlement' was over and that the West could no longer support unsustainable lifestyles, he was not referring to environmental sustainability or to catastrophic climate change.* Nor was he concerned with independence and self-sufficiency – the desire to provide as much of one's needs as possible.

Rather, the then Shadow Treasurer was focused almost entirely on the economy and the unsustainability of welfare payments. It was not so much that the Australian economy needed to cut back on the notion of 'universal entitlement', in order to balance the budget at a time of financial crisis,

---

* Joe Hockey, 'The end of the age of entitlement', text of Shadow Treasurer Joe Hockey's speech to the Institute of economic affairs in London on 17 April 2012, http://www.smh.com.au/national/the-end-of-the-age-of-entitlement-20120419-1x8vj.html, accessed 7 November 2016.

but more to do with Hockey's desire to cut back on welfare to those he called 'leaners' (those who were unworthy of assistance) as opposed to 'lifters' (who kept the economy running smoothly). Prosperity was clearly linked to economic growth and entitlement was for those who participated in the growth economy. For Hockey, the growth model that was central to capitalism would result in independence for the majority of citizens and remove the reliance on welfare. There was, however, a significant minority that sought to cut back for other reasons.

Around the same time as Hockey's landmark speech, Guy McPherson, editor of *Conservation Biology*, also called for an end to the age of entitlement.* But this time, unlike Hockey, it was one that related to over-consumptive lifestyles. Industrial culture and climate change were threatening the very survival of *Homo sapiens*. McPherson's solution was to look for an individual response – something he could personally do about this predicament. His answer was to leave his academic job in the city and 'go back to the land' with his family to live more sustainably – a move that many others had made many times before him.

McPherson's motivations were indicative of self-providers everywhere:

---

* Guy McPherson, 'Going Back to the Land in an Age of Entitlement', *Conservation Biology*, Vol. 25, No. 5, 2011, pp. 855–7.

> The reasons for changing my lifestyle reflect my core beliefs. I could no longer contribute to an empire built on an industrial economy based on consumerism, and thus resist imperialism (i.e., the dominant paradigm, which is characterized by oppression and hierarchy), or live in a city, which is not supported by my moral imperatives.

Self-sufficient homesteading presented a real alternative to the failings of capitalism, albeit one that did not fit with Hockey's philosophy. It was a total rejection of the growth model but one that proponents felt could truly represent their values. It was within the confines of self-sufficiency that true independence and freedom would be possible.

Self-sufficiency has important historical antecedents developing simultaneously with the growth of the consumer market. But it remains a real alternative for those who want to escape the confines of the market and a life based on a string of consumer choices. Self-sufficiency, the act of providing as much of one's needs as possible, offers proponents the chance to remove themselves from the growth economy. The reality is that self-sufficiency doesn't have to be an all-consuming lifestyle in which self-providers remove themselves entirely from society to live the life of hermits in the bush. Rather, self-sufficiency allows us to rethink our relationship with consumerism and materialism and come up with an alternative lifestyle that is of our own making. Whether we choose to live on and from the land or grow vegies in pots on our inner city balconies, we are practicing self-sufficiency.

Every time we reject the growth economy we have a little win; the more we do the greater the victory. Far from merely being another form of green consumerism, self-sufficiency is something that cannot be co-opted by marketers and advertisers and sold back to us in neat, well-designed packages. Self-sufficiency gives us the autonomy to make our own decisions about how we want to live and what impact we are to have on the planet. Importantly, self-sufficiency also removes us from the 'leaners and lifters' dichotomy. We are independent from an economy we have no faith in and that has little faith in us.

This paper is based on the notion that the reliance on materialism for growth and consumer wellbeing is not only a flawed concept but also leads to environmental catastrophe and climate change. Self-sufficiency offers real solutions to these problems.

CHAPTER 1

THE PROBLEM OF GROWTH

In 2012 Australians spent $642 billion on consumables. Of this, $78.4 billion was spent on cars, $14.1 billion on alcohol and $19 billion on recreation. Only $2.2 billion was spent on public transport. Despite its environmental cost, by 2015, 5.3 million of Australians over the age of 14 drank bottled water, an increase on 2014 when 4.9 million Australians drank it.* These are striking statistics for a population of less than 25 million.

Australia, like its counterparts across the developed world, is dependent on growth. Consumer capitalism holds that it is consumption by a nation's citizens that drives this growth and leads to prosperity for all. It is, therefore, a

* Roy Morgan Research, 'Bottled water consumption booming', article No. 6763, 19 April 2016.

good thing that bottled water consumption is on the rise. It is good for the country and its people.

But, if the majority of Australians have benefited from capitalism, why should we reject the growth economy? This growth obsession has, of course, led to environmental destruction and the increasing threat and reality of catastrophic climate change.

Capitalism has been one of the greatest success stories of the past two hundred years. But is capitalism the harbinger of evil that its critics have led us to believe?

Capitalism replaced the mercantile system that operated predominantly between the sixteenth and eighteen centuries. Unlike capitalism, which is largely attributed to the publication of Adam Smith's *The Wealth of Nations* in 1776, the mercantile system was not about appeasing the consumer interest or keeping consumer prices low. Based on the accumulation of gold, governments regulated markets, maximised exports and minimised imports. Often characterised by high prices and product shortages, mercantilism was based on the accumulation of gold to generate wealth.

Modern capitalism, the combination of trade and consumption, was a particularly attractive prospect to those who lived at a time of high levels of poverty. Broadly speaking, our modern reliance on consumer-based trade can be traced to the eighteenth century and the writings of Smith. The changing importance of trade was that it would create wealth for the whole of society. Adam Smith's philosophy of wealth was radical as universal consumption was an ultimate goal. 'This is a democratic, and hence radical, philosophy of wealth', writes Smith,

> Gone is the notion of gold, treasures, kingly hoards; gone the prerogatives of merchants or farmers or working guilds. We are in the modern world, where the flow of goods and services consumed [and earned] by everyone constitutes the ultimate aim and end of economic life.*

It is the linking of commerce and consumption that makes Smith's legacy relevant here. His oft cited adage that 'consumption is the sole end and purpose of all production' is neatly coupled with 'and the interest of the producer ought to be attended to only so far as it may be necessary for promoting that of the consumer'. It was 'perfectly self-evident', wrote Smith, that 'it would be absurd to attempt to prove it'. Yet it is worth stressing that the consumer is elevated under capitalism:

> But in the mercantile system the interest of the consumer is almost constantly sacrificed to that of the producer; and it seems to consider production, and not consumption, as the ultimate end and object of all industry and commerce.†

---

* Adam Smith, *The Wealth of Nations*, Book IV, Camberwell: Penguin, 1776, p. 53.
† Ibid., p. 245.

And so, capitalism was born and the elevation of the consumer begins.

As Duncan K. Foley explains: the pursuit of wealth was now legitimated for everyone and that wealth could be produced anywhere not just on the land;* it was labour now and not nature that was the primary source of value.† The market was to be left alone with competition regulating economic life. Robert Heilbroner, writing in 1953, during the long economic boom of the postwar years, noted:

> And of course Smith is right. If the working of the market is trusted to produce the greatest number of goods at the lowest possible prices, anything that interferes with the market necessarily lowers social welfare.‡

It is a mistake to view Smith as a conservative economist, writes Heilbroner, he was in fact pro-consumer and less enamoured by the new creators of wealth: 'he was more avowedly hostile to the motives of businessmen than are most contemporary liberal economists'.§

---

* Duncan K. Foley, *Adam's Fallacy: A Guide to Economic Theology*, Cambridge (Mass): Belknap Press of Harvard University Press, 2006, p. 49.

† Robert Heilbroner, *The Great Economists: Their Lives and their Conceptions of the World*, London: Eyre and Spottiswoode, 1953, p. 49.

‡ Ibid., p. 70.

§ Ibid.

Smith described the making of the consumer, the mass of citizens that were to become the beneficiaries of private interests engaged in wealth creation. Consumers – that great mass of society – were to be the central beneficiaries of Smith's radical system of capitalism, not through direct government policy, but rather through the unregulated 'invisible hand of the market'. It would be misleading to assume that consumers were central to business decision-making processes during Smith's time like they were to become in the 1950s with the development of marketing.* However, they were the intended beneficiaries of the capitalist system as envisaged by Smith in that an efficient market would deliver benefits to the whole society. For more than two centuries, capitalism became a central focus for both its proponents and opponents as efforts were made to civilize and tame it. However great the actual power of the consumer, it is clear that the consumer – and consumerism – plays a central part in the capitalist philosophy.

Smith described the straightforward nature of his consumerist philosophy thus:

> It is the maxim of every prudent master of a family never to attempt to make at home what it will cost him more to make than to buy. The tailor does not attempt to make his own shoes, but buys

---

* Amanda McLeod, *Abundance: Buying and Selling in Postwar Australia*, Melbourne: Australian Scholarly, 2007.

> them of the shoemaker. The shoemaker does not attempt to make his own clothes, but employs tailor.[*]

This view was extrapolated to the whole of the economy. If it is cheaper for a foreign country to make and supply goods than domestic markets then that is what should happen. It simply made sense to do so.

Under Smith's vision, capitalism, trade, and consumption would benefit everyone. Nevertheless, for as long as there have been proponents of wealth creation there have been ferocious attacks on the status quo. Critiques developed simultaneously as opponents sought to find alternatives.[†] Marx, of course, believed that capitalism would eventually and inevitably collapse, though he did not say much about what a communist society would look like after its demise only that it would be led by the proletariat. Marx foresaw many of the ironies of the capitalist system. Universal abolition of poverty did not occur but instead exacerbated the gulf between the rich and poor. Workers, too, were alienated from their work and exploited for their labour.

Although capitalism has lost some of its shine, it has not lost its prominence in being promoted as the key to

---

* Smith, *The Wealth of Nations*, Book IV, p. 33.

† See for example: Samuel Alexander and Amanda McLeod (eds.), *Simple Living in History: Pioneers of the Deep Future*, Melbourne: Simplicity Institute, 2014.

economic and material prosperity. Its proponents still hold it to be the only way to solve poverty in the developing world, national debt and other economic problems.

Critiques rejecting the growth model, on which prosperity has been based, have been around as long as the model itself. Indeed, it is almost certain that Smith himself would not recognise the realities of capitalism in the twenty-first century. Yet pundits continue to expound its virtues. Critiques, however have been equally strident.

Writing more than two centuries after Smith, philosopher and ethicist Peter Singer explained in 1994 that there are two very good reasons for rejecting materialist prosperity as an economic growth model. On the one hand, there is little evidence to suggest that materialism equals happiness. On the other, that modern life based on mass consumption has had a significantly negative impact on the environment with climate change threatening to destroy life as we know it. Singer lamented that 'We are running up against the limits of our planet's capacity to absorb the wastes produced by our affluent lifestyle'.* It is on this latter problem that is my concern here.

More than twenty years after Singer's call to arms, the need to rethink our relationship with mass consumption has never been more urgent. Yet, we keep on consuming. We keep on consuming despite our increasing indebted-

* Peter Singer (ed.), *Ethics*, Oxford: Oxford University Press, 1994, p. 179.

ness and we keep on consuming at record levels despite the dire environmental predictions and consequences. We keep on consuming and governments continue to pursue the growth model regardless of the harm that results. We keep consuming despite the onset of climate wars.

The climate crisis has been directly driven by capitalism, by over-consumption and the belief that the consumer is king. Prosperity based on material acquisition and the myth of consumer sovereignty has already led to countless environmental disasters and economic downturns. This paper offers a solution to capitalism's discontents by arguing for radically simplified lifestyles in order to move away from out-dated and unworkable notions of consumerism. And the time to do so is now!

Right now.

There are a number of important common assumptions made about climate change. For example, that climate change is real and it is caused by human behaviour.

I am not going to debate this point, there is little point questioning the reality of anthropogenic climate change. It is here and it is impacting the planet now. Catastrophic bushfires, hurricanes, droughts, floods, rising sea levels, and other extreme weather events are becoming the norm.

Climate change can be seen as a 'threat multiplier'. It interacts with pre-existing problems and acts like an accelerant to instability. The causes and consequences of conflict in Africa, the Middle East and Asia can be directly linked to climate change. Climate wars cause disruptions, poverty, mass migration, displacement and civil unrest. We have, as they say, reached the Age of Consequences – catastrophic consequences that are not merely climatic changes. Exacerbating existing problems, competition for scarce resources and poverty culminate in serious tensions. Climate change mitigation has given way to adaptation. Yet, we can no longer rely on adaptation alone. We need to build resilient societies. Self-sufficient societies that can withstand the consequences of climate change.

There are two further assumptions often made about climate change, that governments should be responsible for doing something about climate change; and we should not make individuals feel guilty about what they ought to be doing about climate change.

These two points are interrelated. On the one hand, we should continue to wait for a top-down solution. It is clearly governments that should do something about climate change. On the other, individuals should not feel guilty and need not change their behaviour because individual action is infinitesimal. Individual action, so the conventional argument goes, simply does not make a difference.* However, this view is at odds with the elevation,

* See for example: Walter Sinnott-Armstrong, 'It's Not My Fault: Global Warming and Individual Moral Obligations', chapter 18 in Stephen M. Gardiner (et al), *Climate Ethics: Essential Readings*,

over the last sixty years, of the consumer. But where is the notion of consumer autonomy and sovereignty in this equation? Where is the notion of responsibility and accountability?

Oxford: Oxford University Press, 2010, pp. 332–46.

CHAPTER 2

CONSUMER POWER

According to governments and marketers, consumers have significant power in the marketplace. They have the power to choose to buy one product over another, the power to switch brands, and ignore marketing's messages. Australian consumers, like their counterparts across the developed world, have almost unlimited choices. However, more choice doesn't always mean better choice. Indeed, many of our choices may have unwanted consequences not the least over-consumption and climate change.

Since the end of the Second World War, the pursuit of economic prosperity and security has been focused primarily on material acquisition and increasing consumer choice; most Australians reaped the economic benefits. With consumption as the key to prosperity, all efforts were geared to making the market work efficiently and equitably. The commercial sector, the government, and the independent consumer movement all pursued this aim through the various measures that formed the consumer policy framework and established the notion of consumer rights.

At the height of the consumer revolution in 1962, U.S. President John F. Kennedy introduced his four consumer rights to Congress: the rights to be informed, to safety, to be heard, and to choose which largely determined the direction of consumer protection and advocacy. In 1985, the United Nations extended Kennedy's notion of consumer rights by adding: the satisfaction of basic needs, to redress, to consumer education, and to a healthy environment. This final right – the right to a healthy environment – has been mainly applied to the elimination of pollution in developing countries and has not been prioritised during times of heightened climate change. Neither has the right to the satisfaction of basic needs been the primary focus of consumerism and materialism. Overwhelmingly it has been the 'right to choose' that has been the focus of marketers and policy makers and has ultimately led to unfettered consumption. Governments depend on consumer spending for economic prosperity. Consumers, it is argued, can help to buy their way out of economic recessions.

More recently, the consumer policy framework has been designed to empower consumers and, in turn, stimulate competition. 'Effective competition', the Productivity Commission argued, 'is stimulated by empowered consumers and responsive suppliers that trade fairly'.* The Commission's 2008 *Review of Australia's Consumer Policy Framework*, which led to the new *Australian Consumer Law* 2011, recognised the importance of competition policy as being the greatest driver for improving the wellbeing of Australians. Consumer choice was central to its delivery:

> Most notably, reductions in trade barriers and competition policy reforms have put downward pressure on prices, enhanced product quality and increased consumer choice. Indeed, almost all economic policies are ultimately aimed at improving consumer wellbeing.†

While the Commission was careful 'not to downplay the importance of consumer's rights, which for many are the starting point for assessing a desirable policy framework', the broader body rights of consumers could not always

---

* As federal treasurer, Peter Costello instructed the Productivity Commission in 2006 to examine Australia's consumer policy framework with a view to introducing a 'single generic consumer law' which would apply across all Australian jurisdictions (the *Australian Consumer Law* came into effect in January 2011).

† Productivity Commission, *Review of Australia's Policy Framework*, Final Report, 2008, p. 2.

be the ultimate goal. Nor, from the Commission's point of view, were they always in the ultimate interests of the consumer body as a whole. The Productivity Commission argued that 'while broadening those rights may be in the interests of the wider community, the associated costs must always be considered as part of the policy formulation process'.*

The current consumer policy framework as it has been historically structured is largely unequipped to consider issues that extend beyond individual consumer participation. Governments, consumers, and businesses are so caught up prioritising choice and promoting acquisition that other, often more important, issues do not get examined.

So why has the government not pursued the path of sustainable or ethical consumption for individual consumers? Put simply, lifestyles based on frugality, austerity, and thrift are not seen to be easy politically. Individual choice is the catch cry of the market with the consumer freedom to choose equated with democratic freedoms. Australians, like their American and European counterparts, view austerity negatively as a restriction of rights.

It is within this context that the Rudd Labor Government introduced its Economic Stimulus Package during the Global Financial Crisis in 2009. Instead of calling on citizens to exercise economic restraint, to save rather than

---

* Ibid., p. 12.

spend to weather the economic storm, the Rudd Government sent us out to spend as if our (economic) life depended on it. This measure was largely successful in economic terms. It initiated stimulus spending to avoid economic recession and a potential economic depression with a $42b Economic Stimulus Plan. Success of the Government's Stimulus Package reinforced the message of consumption and its importance in creating economic wellbeing.

Despite significant spending on infrastructure and education, a major part of the package – $12.2b – was in the form of 'bonus payments' to encourage individual consumer purchases. The message was explicit: good citizens should not spend the money paying off debt and they certainly should not save it. The good citizen was not to tighten his/her belt and exercise thrift in a time of economic hardship; the good citizen should go forth and spend. And, spend we did.

Whatever the root causes of Australia's successful negotiation of the Global Financial Crisis, Labor's actions reinforced that consumption was a democratic duty and that a good citizen was a good consumer (and vice versa). This presents a paradox for those charged with addressing the problems and consequences associated with over-consumption – waste, environmental damage, global warming, climate change, etc. Whose responsibility is it to deal with these problems and how can these aspects be reconciled? It is worth investigating possible alternatives.

Labor's intervention contained significant incentives to encourage individual consumption. But the message of spend, spend, spend as if the prosperity of the country depended upon it was as problematic as it was compelling. This narrow definition of citizenship – good citizen as good consumer – has wider ramifications for a healthy democracy (one that is unable to meet challenges and spread rewards widely and evenly).

While the Stimulus Package included spending on what can be labelled as environmentally sustainable initiatives, including rebates for energy efficiency, as well as the ill-fated housing insulation program, individual consumers were not encouraged to spend their individual bonus payments in environmentally sustainable or ethical ways. In so doing, the Government missed an important opportunity to better sell an environmental sustainability or austerity message.

Governments, for fear of being criticised for reining in rights, have been reluctant to rein-in consumption. But the benefits for the environment of moderate consumption levels are obvious. We have to, however, be careful that we are not merely offered superficial alternatives. We need substantial and permanent change as well as changing light globes, installing water tanks and solar panels and using

E10 petrol. We need to use less power, water, and petrol *per se*.

We need to live simply. We need to exercise our power to say no – to capitalism and to the growth model. We need to take responsibility for our part and do our bit.

CHAPTER 3

THE CLIMATE CRISIS

The climate is under siege. Australia's worst bushfire disaster, the Black Saturday bushfires – in February 2009 – was the result of land clearing, climate change, high temperatures and drought. The event killed more than 170 people. The fires did untold damage to houses, property and native animals and livestock.

Extreme weather events are becoming increasingly common in Russia, and the 2012 drought – following hard on the severe drought of 2010 – confirmed this trend. During 2012, 22 regions suffered crop losses, with a state of emergency declared in 20 of these. The flow-on effects were disastrous.

There is a direct link between Russia banning wheat exports as a reult of the drought and the Arab Spring. The

Arab Spring or Democracy Spring was a revolutionary wave of both violent and non-violent demonstrations, protests, riots, coups and civil wars in North Africa and the Middle East. Hunger exacerbated existing tensions. Protesters took to the streets across the Arab world in 2011, pushing their leaders to end decades of oppression. The Middle East and North Africa were engulfed in an unprecedented outburst of popular protests and demand for reform. It began in Tunisia and spread within weeks to Egypt, Yemen, Bahrain, Libya and Syria. While the causes of the Arab Spring are complex, climate change has been called a facilitator of tension.

In August 2013 floodwaters inundated up to one fifth of Pakistan and affected an estimated 20 million people. The list of climate events and its consequences are almost endless.

Each year is marked by unprecedented temperature increases. 2015 was a year of climate records. In 2016 another record was set. The Great Barrier Reef, for example, has experienced two unprecedented events of coral bleaching to more than two thirds of it. It remains unclear whether the Reef will recover.

The effects of climate change are more than frightening. Money is unlikely to protect us from its consequences. Adaptation may be all we have.

The climate cannot withstand the effects of anthropogenic climate change. It cannot prevent the Adani coalmine from destroying what is left of the Great Barrier Reef. Climate change continues to change life as we know it. Without swift, universal and bipartisan government action, we may only have our own resources to rely upon. Government action does not seem forthcoming.

During the United Nations climate change conference in Copenhagen in 2009 Prime Minister Kevin Rudd called climate change 'the greatest moral, economic and social challenge of our time'. Despite almost universal agreement on the urgency to combat climate change, Australia and other major economies, have done little of substance to fight it. Indeed, Australia's political action on climate change could be called one of the greatest failures of our time.

Although its findings are generally considered conservative, the Intergovernmental Panel on Climate Change's view is unambiguous. In the period since the end of the Industrial Revolution significant and catastrophic changes have occurred: 'Each of the last three decades has been successively warmer at the Earth's surface than any preceding decade since 1850.' But it is since 1950 – the beginning of the postwar consumer boom – that most damage has been done. The Panel reported in 2014 that:

> Warming of the climate system is unequivocal, and since the 1950s, many of the observed chang-

> es are unprecedented over decades to millennia. The atmosphere and ocean have warmed, the amounts of snow and ice have diminished, and sea level has risen.*

Regardless of the significant groundswell amongst well-meaning urbanites in developed nations, the answers to one of the greatest questions of our time are not to be found in the stalls of farmers' markets, in the pages of so-called green magazines, or in heated discussions around kitchen tables, however virtuous these things might be. While we wait for top-down solutions, for governments to take climate change seriously rather than merely paying lip-service to it, we should also do our bit: by growing our own vegies, keeping bees, and sewing our own clothes. We should generate our own energy, ride bikes, and take public transport. We should not fly, eat red meat (unless we produce it ourselves), or factory farmed produce. We should only eat sustainable seafood. We really should do our bit for the planet. Very seriously the list of things we should and shouldn't do in the name of climate change is almost endless.

But time is clearly up. We simply cannot wait any longer for a serious political commitment, however we may need it. As citizen-consumers we have the power to do some-

* IPCC, 2014: Synthesis Report. Contribution of Working Groups I, II and III to the Fifth Assessment Report of the Intergovernmental Panel on Climate Change.

thing about climate change – we should take responsibility for our hand in it.

Given past unsuccessful attempts to broker a solution we cannot rely on governments to implement strident policy initiatives. But because governments can easily override the political decisions of previous administrations, there is heightened need for individuals to take action. Philosopher Holly Lawford-Smith argues that we need bottom-up action to combat climate change, because it is simply too easy for governments to repeal top-down solutions. We should, instead, focus on individuals and their great potential power, she writes:

> I think if we stop talking about what states ought to do about climate change, and start focusing on what individuals ought to do, we might find that states end up being able to do what they ought to do as a consequence.*

One way of radically reducing consumption is to place formal restrictions on consumer rights. While 'free choice' may be central to commercial decision-making, there is a good argument for restricting our rights as consumers. Governments must enable us as citizens to step back from the consumer market just as governments must move away from the growth model of material acquisition.

---

* Holly Lawford-Smith, 'Difference-making and Individuals' Climate-Related Obligations', final draft as at 29th July 2014, p. 21.

While free choice is the mantra of the market, only a specific aspect of this is promoted – the free choice to choose – to say 'yes' to consumption. Yet, saying no must be prioritised instead. We must transition quickly to self-sufficiency in order to embrace a simpler way of living.

Nor can we merely buy our way into green solutions to fix climate change. Consuming green is still based on conventional economics. In fact green solutions are too easily co-opted by marketers and sold back to us as consumers with an increased price-tag. As journalist and activist Naomi Klein writes:

> Consuming green just means substituting one power source for another, or one model of consumer goods for a more efficient one. The reasons we have placed all of our eggs in the green tech and green efficiency basket is precisely because these changes are safely within market logic – indeed, they encourage us to go out and buy more new, efficient, green cars and washing machines.*

However, self-sufficiency is about far more than consuming green. Self-sufficiency is about taking control of our own consumption practices, reducing consumption and making do.

If we have indeed reached the tipping point on runaway climate change and there is much to suggest that we have,

* Naomi Klein, *This Changes Everything: Capitalism vs. the Climate*, UK: Penguin, 2014, p. 90.

then the solution does not lie in sustainable capitalism. There will be no green solution for marketers to hijack, no technological answers to hope for. We need to transition to simple living, not only for economic reasons but to appease the climate crisis.

Yet, it goes without saying that economic austerity is politically unpopular. However, what I am suggesting need not be a Greece-type bailout. It has little to do with government deficits, although it will improve the government's bottom line. It has little to do with raising taxes, although it will decrease reliance on government coffers. We must all simplify if we are to meet the demands of the climate crisis.

Because of the climate emergency and the lack of political will to do anything about it, self-sufficiency will not only be the choice of the few but necessary for the many. It is as fulfilling as it is empowering. It may be all we have to live a life with meaning in a time of unprecedented upheaval and change. It will allow us to be prepared but also allow us to live purposely.

We need to adapt but we cannot rely on conventional structures to help us to do so. Nor can we rely on geoengineering to change the climate to make the planet more hospitable.

The vast geoengineering programs designed to mitigate global warming's negative consequences include absurd mass technological solutions such as liming the oceans to mitigate acidification, biological sequestration, regulating sunlight (solar radiation management), cloud brightening to reflect solar radiation away before it reaches the earth and even eliminating cirrus clouds. Instead of carbon emission reduction by stopping or reducing the burning of fossil fuels, geoengineering refers to technological programs designed to alter the climate to mitigate the effects of global warming.

Brought about because nations, for a myriad of reasons, have not acted on global warming, geoengineering appears to be a white knight. It is, for its advocates, both a solution and an opportunity not to move away from consumerist lifestyles that rely so heavily on the burning of fossil fuels. But it is no white knight. Many of the consequences of geoengineering are unknown and untested.

In order to live simply we must rethink our relationship with the consumer market and its spoils. By embracing austerity, by rethinking the meaning and relationship between luxuries and necessities we can do our bit to mitigate and adapt to the climate catastrophe that is upon us. We can all simplify regardless of our economic means. Individual responses such as simple living and self-sufficiency are something we can all embrace. But until we do we, and the planet, seem doomed.

Many have extolled the virtues of downsizing lifestyles,

to work less and to have more leisure time. While the self-sufficient life decreases our reliance on employment income and enables us to become more independent, increased leisure time is not the goal. In fact, self-sufficiency is about the satisfaction that comes from hard work. The more self-sufficient you are and the freer you are from outside pressures, the harder you will work. But, that is the whole point. Rather than getting satisfaction through consumption alone, the self-provider gets satisfaction from both production and consumption, from doing it themselves. While downshifters may value having more time to 'stop and smell the roses', the self-provider is far more likely to pull out the roses and grow tomatoes instead.

Countless books have been written on the climate emergency. Some have suggested downsizing lifestyles to combat the alienation caused by consumer capitalism. Others have suggested that downsizing will be the panacea to the meaningless banality of lives based on consumerism. However, if downshifting is only about increased leisure time, it is in danger of becoming but another form of consumption that can be co-opted by commerce.

Put simply, the reason for climate change is our exponentially increasing use of fossil fuels. If we have contributed to the burning of fossil fuels through our consumption we have a responsibility to mitigate the effects of climate change. On the one hand, we must mitigate the effects that we have directly contributed to. On the other, we must also change our future behavior so that we do not cause more harm.

It is a long and complex philosophical debate whether our individual consumption actually makes a difference.* When it comes to the link between consumption and the burning of fossil fuels, individual acts might be infinitesimal. It has been argued that when it comes to individual acts (a one-off drive to work in your car or a drink from a plastic bottle of water) our behavior might not make a difference to overall carbon emmissions. If my own consumption doesn't make a difference, it seems it doesn't matter what I do. I don't have to curtail my consumption in order to prevent harm to the climate.

But it makes no sense to calculate the impact of our behavior based on an individual event. We are, after all, lifelong consumers. However, when we consider a pattern of behavior – or a lifetime of consumption – our individual actions might make a significant difference. It is our consumptive lifestyle as a whole that makes sense to measure. And it might matter a whole lot.

It actually might make such a difference that we can calculate the harm that we cause. By calculating our carbon footprint we can get an idea about how much impact

---

* Shelly Kagan, 'Do I make a Difference?', *Philosophy and Public Affairs*, 1 April 2011, Vol. 39, No. 2, pp. 105–41.

our consumption is having on the environment. Once we know the extent of our carbon emissions then we will be able to offset our consumption.

Offsetting is one way we can mitigate or eliminate the harm that we cause through our consumptive decisions. Used primarily for offsetting air and motor vehicle travel and energy consumption, offsetting actually does neutralise individual harms.

But buying carbon offsets is not the way to go either. There are a number of inherent problems with offsetting, including the fact that it is difficult to assess the efficacy of various offsetting schemes. One may choose to offset the carbon emissions of a return trip from Melbourne to London by buying carbon offsets from an organisation that plants trees in developing countries. While it makes sense to offset travel emissions, it is less easy to offset other forms of consumption – particularly for seemingly small purchases. How many trees do we have to plant, for example, to mitigate the effects of bottled water consumption, or a plastic punnet of strawberries, or even the use of single plastic bag?

There are computer programs that calculate the global footprint of consumers. But the programs do not calculate all of our consumption; rather they provide a rough guide of how many tonnes of carbon we emit from the petrol, electricity and gas we use. We can, of course, over-estimate the amount of carbon and then offset that. But, we are not getting the true global footprint of our actual consumption. It does not take into consideration how much carbon we emit when we purchase consumer durables (including white goods and computers and other technology), food or entertainment and many, if not most, other products.

The problem with offsetting is that it only works if *everyone* does it. Unless everyone – including all individuals and manufacturers the world over – offset all of their energy consumption there will be no way to capture all carbon emissions and account for all embedded energy in the products we consume.

While we may have an ethical obligation to offset our carbon emissions, and everyone else shares this ethical obligation, it is unlikely that everyone will actually participate in offsetting schemes.

However, there is another aspect of offsetting that makes me uncomfortable. While the actual harm we cause might be extinguished by offsetting (if everyone does it) we need to change our consumer behavior entirely. Offsetting does nothing to encourage this. Indeed, it extinguishes any responsibility to consume sustainably. If we offset our energy use, for example, there is nothing stopping us from using as much petrol, gas and electricity from fossil fuels, as we want. If we offset there is no actual limit to what we can (ethically) consume.

However, whether or not offsetting works, these are, as Schumacher told us, finite resources that cannot be replaced. We may cause harm by contributing to further resource depletion. There are a number of products that *need* to be produced – including medications, renewable energy infrastructure, etc., – using oil and its by-products (including petro-chemicals). Renewable energy will not be able to replace oil if we want to continue on 'business as usual'. We need to reduce consumption and adopt a new economic system. While there are many reasons to reduce consumption of fossil fuels beyond their finite nature, they remain necessary for a small number of consumable products.

Degrowth 'means a phase of planned and equitable economic contraction (rather than growth) in the richest nations, eventually reaching a steady state that operates within Earth's biophysical limits'.* It has been argued that degrowth is a political, economic and social movement based on ecological economics, anti-consumerist and anti-capitalist ideas.

It is the idea that overconsumption lies at the root of environmental problems but also that reducing con-

* Samuel Alexander, 'Life in a Degrowth Economy and why you might actually enjoy it', *Conversation*, 2 October 2014 https://theconversation.com/life-in-a-degrowth-economy-and-why-you-might-actually-enjoy-it-32224, accessed 2 June 2016.

sumption does not harm individual wellbeing and destroy prosperity. Rather it has many benefits including living within the earth's limits. Self-sufficiency is entirely compatible with the degrowth ethic. Rather than focus on the national economy, self-sufficiency is designed to work at the household level – it is something that individuals can do themselves rather than wait for a top-down solution to their problems.

I have argued that consumption by individuals makes a difference and that offsetting cannot mitigate the harm to the environment caused by personal consumption because it is simply too difficult to calculate. If we should take responsibility for the harm that we cause what are we to do? Self-sufficiency offers us answers to some of these problems.

What we need to do is lessen our impact on the environment. Self-sufficiency does that. If we cannot offset our carbon emissions that result from our purchases we should avoid certain products altogether. There is simply nothing endearing about bottled water, about plastic bags and disposable coffee cups. Self-sufficiency lessens our reliance on traditional consumption; it also cuts waste.

Maybe self-sufficiency should be seen as a form of offsetting, because through it we can mitigate harm. Self-sufficiency advocates adopting carbon neutral activities; activities that won't harm the earth. Some, like gardening, may even drawdown carbon by sequestering it in the soil. Large-scale industrial agriculture that relies on the use of petro-chemicals, deforestation and mono-cropping, is not known for carbon sequestration. Indeed, such farming practices are responsible for massive carbon emissions.

But self-sufficiency is about a different type of farming altogether.

CHAPTER 4

SELF-SUFFICIENCY – WHAT IS IT?

Now self-sufficiency is not 'going back' to some idealized past in which people grubbed for their food with implements and burned each other with witchcraft. It is going *forward* to a new and better sort of life, a life which is more fun than the over-specialized round of office or factory, a life that brings challenge and the use of daily initiative back to work, and variety, and occasional great success and occasional abysmal failure. It means the acceptance of complete responsibility for what you do or what you do not do and one of its greatest rewards is the joy that comes from seeing each job right through – from sowing your own wheat to eating your own bread, from planting a field of pig food to slicing a side of bacon.*

* Seymour, *The Complete Book of Self-Sufficiency*, p. 7.

In this chapter I outline the historical and philosophical dimensions of self-sufficiency. Put simply, self-sufficiency means doing as much for oneself as possible. In an age of boundless consumption, self-sufficiency is the ultimate act of rebellion. Self-sufficiency enables practitioners the chance to remove themselves almost entirely from the consumer market. Indeed, this is the ultimate aim of self-sufficiency – to live a life outside the confines of materialism and mass consumption.

The notion of self-sufficiency is complex and, as Allaby and Bunyard suggest, could be confined to a discussion on 'how to grow vegetables and whether to keep goats or ban them from the face of the earth, given their propensity for debarking one's favourite fruit trees'. Yet, as they more seriously observe, there are more important issues at play including 'political structures, man's future, freedom and environmental degradation'.*

Post-industrial self-sufficiency is a particularly modern phenomenon. As an alternative to capitalism and consumerism, self-sufficiency developed simultaneously with the development of the consumer market. Ultimately, of course, the answer to the problem of dependence was self-reliance. The method by which to be self-reliant was through practicing and gaining the skills of self-sufficiency. The majority of those promoting the philosophy of self-suf-

---

* Michael Allaby and Peter Bunyard, *The Politics of Self-Sufficiency*, Oxford: Oxford University Press, 1980, p. 32.

ficiency were those who had already made the shift: Henry David Thoreau, Ralph Borsodi and Helen and Scott Nearing, were the oft-cited pioneers. Yet, different forces separated the two eras of those who found the solution to the problem of capitalism in going-back-to-the-land. While early proponents such as Borsodi and the Nearings were driven by an attack on production, later converts targeted the problem of consumption. The notion of self-sufficiency that emerged with the lived experiments of the Nearings and Borsodi in the first half of the twentieth century was driven by economics in hard times – self-sufficiency was a way out of economic depression.*

When John Seymour released his *Complete Guide to Self-Sufficiency*, from which the quote at the start of this chapter is drawn, the book quickly became, and remains, the bible of self-sufficiency. Seymour prided himself on the fact that he did not need rely on the consumer market to provide for his family's needs. His various lifestyle experiments in the 1960s and 70s in England and Wales, on which he and his wife Sally both worked, became the ideal of many who sought to be more independent and in control of their own destinies, rather than at the mercy of the vagaries of the consumer market.

By the 1960s and 70s an important shift had occurred.

---

* Amanda McLeod, 'The Nearings', in Samuel Alexander and Amanda McLeod (eds.), *Simple Living in History: Pioneers of the Deep Future*, Melbourne: Simplicity Institute, 2014, pp. 139–48.

Self-sufficiency by this time was a strident response to affluence and consumerism. Indeed, self-sufficiency during this period can be understood as self-sufficiency in a time of plenty.* As the twentieth century progressed into the 1980s and beyond, self-sufficiency was attractive again to those who were living in economically hard times. By the beginning of the twentieth-first century a stronger environmental focus would emerge.

Self-sufficiency has always had an environmental ethic driving the movement, particularly as a response to pollution and other harms associated with industrialisation. In the 1970s, for example, the fear that fossil fuels – particularly oil – would run out was central to the movement during this time; everyone, would be forced to live more self-reliantly. This apocalyptic scenario driven by the 1970's oil crisis certainly influenced many proponents of self-sufficiency.† Some have argued that we will be forced to live more self-sufficiently due to peak oil. Peak oil (the point where the maximum rate of extraction is reached, after which it will be in terminal decline) certainly puts conventional economics at risk. But we can't wait for the oil to run out. It is climate change that now makes an

---

* Amanda McLeod, 'Self-Sufficiency in a Time of Plenty: Mass Consumerism and Freedom in 1970s Australia', *History Australia*, Vol. 14, No. 3, September 2017, pp. 1–19.

† John Seymour, *The Complete Book of Self-Sufficiency*, London: Faber and Faber, 1976.

alternative economic system necessary.

By the beginning of the twenty-first century, an even more strident environmentalist critique would also run through the self-sufficiency movement. By this time, industrialisation, materialism and consumerism were clearly recognised as the cause of climate change and self-sufficiency was widely seen as the solution to these problems.

Put simply, self-sufficiency means to be self-supporting or needing no outside help in satisfying one's basic needs. But it is more than that. Self-sufficiency is specifically a way of removing oneself from the consumer market to provide for oneself as much as possible. It means growing and processing your own food, animal husbandry, crafting, making clothes, generating energy, and even building your own home. You might choose to do all these things or only some. It can mean living in the country or the bush away from the city and all its modern conveniences. Or it might mean living in the city but removing oneself from conventional production processes. You might only be able to grow food on your balcony or you might be lucky enough to turn your suburban backyard into a food forest with chickens and ducks.

Self-sufficiency means doing-it-yourself rather than buying into so-called 'green' consumption. Ultimately, self-sufficiency is not about buying at all. It is about making, making do and doing without. It is the search for a better kind of lifestyle in which the individual negotiates her own destiny rather than appropriating one that has been designed by marketers.

Self-sufficiency does not mean you cannot earn money, only that, when it comes to spending we should spend our money ethically and in ways that enhance the self-sufficient life. Reducing our reliance on traditional capitalism may mean 'going off grid' and generating your own electricity. This, of course, requires money. The investment, however, pays a different kind of dividend, one that goes beyond reducing electricity bills. Importantly, it reduces damage to the environment and reliance on the damaging effect of fossil fuels.

There are many advantages of a self-reliant life. The money you will save can be spent on more profitable pursuits – perhaps you will be able to work less and live more. Maybe you will be able to spend the savings on infrastructure to help you live more independently. It is worth remembering that a self-sufficient life costs what you are willing to spend on it.

Being self-sufficient means many different things depending on your needs and wants. You will save money and be more independent. You can take a step back from the consumer market. You can be an ethical consumer. You will reduce waste and lessen your impact on the environment. You can do something positive for the planet. And, enjoy the satisfaction of being able to do it yourself.

Rather than relying on conventional factory farming practices, lessening our reliance and impact on the environment may mean raising our own animals. Moreover, if we do it ourselves we are able to control the outcome of production and consumption processes. We are able to prevent further environmental destruction and climate harm. For example, by buying second-hand goods we are not responsible for unethical production at all. We can build dwellings for ourselves and for our animals. We can make our own clothes and grow our own food without reliance on third parties that exploit workers and ecosystems. The more we do to be self-sufficient the more we are able to remove ourselves from unethical production processes. Despite hard work there are few disadvantages of 'doing it ourselves'.

But hard work is not the same as drudgery – which suggests hard work without reward. Within the self-sufficient context, hard work is associated with satisfaction and reaping the fruits of your labours. As John Seymour reminds us, self-sufficiency is not about going back to some idealised past but going forward to a new and rewarding future.

Self-sufficiency offers proponents the opportunity to be in control of their own lives. But self-sufficiency is also about living lightly on the planet. By becoming the producer of one's own needs we are able to make sure that our processes are environmentally sound. We have control over where our goods come from and where they will end up when we are finished with them.

There is a strong desire, within self-sufficiency, to consume and produce goods ethically. It is as a direct response to modern consumer processes that self-sufficiency is so desirable. It offers us the chance to reject the system that is directly responsible for climate change. By taking control of our own behaviour we are no longer directly responsible for the negative consequences that modern consumption produces.

Self-sufficiency has always been a reaction and a solution to economic uncertainty. In the 1930s, American self-providers sought independence, 'homesteading' on the land at a time of severe economic depression. But self-sufficiency is not just to be pursued in economically lean times.

In fact, the greatest shift occurred in the 1960s and 1970s during the affluence of the long boom. For many, materialism had simply failed to deliver the promise of universal equality and prosperity. Alienation drove the desire to embrace self-sufficiency on the land and, by the 1980s, in suburban backyards. Alienation – as Karl Marx predicted – was the key driver for those who were in the grip of the capitalist system. It is a simple equation: self-sufficiency - capitalism = freedom. And so, across the western world, thousands of proponents chose to 'drop out' and live self-sufficiently.

Self-sufficiency reunites the dual spheres of production and consumption in which the proponent becomes both producer and consumer. This ultimately gives people more power than as mere consumers under capitalism. They have control over their own lives by having control over production.

In the current economic climate of high house prices, and talk of a 'housing bubble', self-sufficiency offers an alternative to mainstream consumer choices. It allows us to prioritise our needs. We are able to downsize our wants before we upsize our lifestyles. Self-sufficiency allows, even requires, us to separate our needs from our wants. Mass consumption and marketing exploits our desire to accumulate what we want. Even if marketing and advertising doesn't create wants (which they probably do) we can control those desires with self-sufficiency.

Self-sufficiency allows us to assess what is really important to us. It is not as if we can only have what we need and not have what we want. This is too simplistic. Rather, if we want something that is not considered a necessity, we can make it ourselves. This prevents unconscious consumption. We are forced to think about our needs and wants and decide between luxuries and necessities. We must make a conscious decision about whether we need to buy two cars or one, or one goat or two.

Capitalism's growth model is built on an ever-increasing demand for an ever-increasing array of products. The notion of 'enough' is not part of the equation. Yet, under self-sufficiency we prioritise different things.

There are, of course, many other benefits that flow from being self-sufficient. By doing-it-ourselves we do not contribute to other unethical practices such as buying clothing that has been produced in near slave – like conditions in the third world. We don't support the unethical treatment of plants or animals produced by industrial or intensive farming. By having control over production processes, we do not have to build our houses from materials such as unsustainable timber harvested from rainforests. How much you do to lighten your footprint is up to you – but the goal of self-sufficiency remains. Self-sufficiency is about consuming and producing ethically.

Although self-sufficiency solves a number of important economic problems such as over-consumption, it also provides answers to the question of unethical consumption. When it comes to unethical consumption, self-sufficiency gives us a framework to say 'no' to consumption *per se*. But there are some things we have to buy – actual necessities that make our lives better. Depending on your lifestyle, your geographical location and other demographic indicators, some goods will be necessary. I am not suggesting that every kitchen gadget is a necessity however much you like to cook. You obviously don't need a new television in every room. However, I am suggesting that depending on where you live, a car might be absolutely necessary to get you to work, the medical centre or the vets. This of course raises the question of what do we do when we need to buy.

There are simply some things that money can buy – some things that we can't make ourselves, cars are one and refrigerators are another. We might lack the skills to do it ourselves but some things are necessary for the self-sufficient life. There is an important principle that can help to guide us when it comes to buying new – buy the best you can afford.

Similarly, others will argue why make it ourselves when we can buy it for less. Self-sufficiency is about a different type of lifestyle – not necessarily a cheap one. It certainly might be cheaper to buy your vegetables from the supermarket than it is to grow them yourself given the time you put in. However, why compare home-grown organic vegetables with those that are conventionally produced? If we grow it ourselves, rather than buy products that are produced on a mass scale, we can control how we grow. When it comes to animals, we can control what we feed them and how they are kept, instead of being reliant on questionable conventional processes. When we do it ourselves we can grow, harvest and produce real organic food – as opposed to industrial organics. We can consume food that is in season, is fresh and what we really like, need and want.

The best option, of course, might be doing without. Consider whether you actually need the product you are going to buy. This might seem obvious but reassessing our needs and wants allows us to take that very important step back from the consumer market. This is central to the self-sufficiency ethic. Buying second-hand is a logical option for the self-provider – if one is aware of the pitfalls. Buying second-hand isn't necessarily the best option (we still need to exercise care and buy the best quality we can afford).

A word of caution. While buying second-hand might remove us from contributing to unethical production, it doesn't necessarily excuse us from unethical consumption. Modern synthetic fabrics, for example, that are derived from petrochemicals, are known to break down and contribute to the problem of plastic pollution. Polar fleece, taking only one product, loses tiny plastic particles every time it is washed. So even buying second-hand products is problematic and far from being straightforward. This is why we need to think about all of our consumer behaviour. We need to think about the consequences of our choices. Maybe the only ethical option will be to buy second-hand clothing made from natural fibres. If we can't find an ethical alternative then we must choose to go without.

Self-sufficiency offers us a unique way of solving problems beyond the home. Self-sufficiency relies on a particular view of consumption that can be described by reference to the adage: 'use it up, wear it out, make it do or do without'. Self-sufficiency is based on an ethic of thrift and recycling. If you can make it yourself that is what you do. If you can do without something – that is the ideal. Make do or do without is the driving force behind the self-reliance philosophy. You simply don't need the latest car, phone or kitchen gadget. All you need is enough 'stuff' to make a life worth living. Some of these things might be bought new; others might be second-hand or borrowed. The aim of the self-sufficiency game is to reduce our consumption, to reject unethical products, reuse what we can, recycle what we can't and rethink consumption. When it comes to self-sufficiency we are independent first and consumers rarely.

A better quality of life means better quality food and other consumables. But it is more than this. Self-sufficiency is about making a better life – about being in there up to your elbows in hard work. It is about ethical and thoughtful consumption. Self-sufficiency is about simplifying but it is not about having an easy life. It is, as Sally

Gordon stated, for

> those people who really love this way of life; who actively enjoy cleaning out a smelly goat shed, or going out to feed the pigs and shut in the chickens in all weathers; who find satisfaction in digging over a rough patch of land or spending hours up a ladder pruning a neglected apple tree.*

This is the essence of self-sufficiency.

---

* Sally Gordon, *Australia and New Zealand Complete Self-Sufficiency Handbook*, Sydney: Australian and New Zealand Book Company, p. 6.

CHAPTER 5

SELF-SUFFICIENCY–IN PRACTICE

Megg Miller told the readers *Grass Roots* magazine in 1972 that there was a better way of life than the one that could be found in the city and the suburbs. Such a life was to be found

> away in the country with plenty of fresh air, sunshine, home-grown fruit and vegetables, and home-baked bread. It's sitting by the open fire in winter eating the food you preserved in summer, eggs from your own chickens, milk from your own cows or goats. For some it's the satisfaction of building their own life, and maybe their own house, without having to rely on a boss, or pay a lot of money to tradesmen, fishermen, canning factories, butchers, clothing makers or bakeries.*

* *Earth Garden*, No. 1, 1972, pp. 2–3.

While Miller was writing more than 45 years ago, the same remains true for those who seek a self-reliant life today. Self-sufficiency is all about doing it yourself. If you can grow it, sew it or build it – that is what you do (you don't outsource it). As Miller explained further:

> If you want fresh vegetables you have to grow them yourself. To get bread with any flavour you have to bake it yourself. To get a quality jumper you have to spin and knit it yourself.*

If you do it yourself you are self-sufficient.

What does a day in the life of a self-provider look like? Busy! But no day is the same. Depending on how many animals you have to tend to, which requires regularity, the self-sufficient life is not necessarily governed by routine. However, there are always things and chores to do. On Monday there are tomatoes to pick and sort in to various forms of ripeness. On Tuesday it is time to process the tomatoes to get them ready for Wednesday when you bottle them in your Fowlers Vacola outfit. You will have to repeat the process multiple times (but not all on the same day) if you want to save enough until next year's harvest. On Wednesday you also need to shell the beans you have been drying and plant the garlic. Thursday you will make bread and freeze peas and carrots which brings you to Fri-

* Reprinted in Meg and David Miller, *Grass Roots the Early Years*, Shepparton: Grass Roots, 1979, p. 7.

day when you will need to can the peaches, make some jam and prune the raspberry canes. On the weekend it is time to muck out the goat shed. And on Monday it is off to work you go again to pick and sort tomatoes.

Then there are seeds to plant and seedlings to grow. There are vegetables to harvest and prepare (and eat!). There are fruit trees to prune and fences and animals that need your attention. You might need to shear your sheep, spin the wool and knit a jumper. And, if you are so inclined, home repairs to do, home butchering to organise and goat hooves to trim. The more you do is up to you.

You might make and cut your own fuel or make your own cheese and yoghurt. You might generate your own energy. You might want to collect your own rainwater, salt beef and make your own pottery. You could also pickle onions, plant corn or can fruit and vegetables. You might even sew your own clothes and make hay while the sun shines. You can do it all or just a few of these things; what you do is up to you. You can live a low cost, high style life; you can design your own home or even a chook palace.

You don't need to own your own home to be partially self-sufficient. You can grow in pots, can vegetables and make your own preserves and crafts. There are so many things to make and do. There is no need to buy loads of books to help you on your path to self-sufficiency. Libraries are full of how-to manuals to help you skill-up while you scale down. You can be the designer of your own life as well as the painter and decorator of your home.

For decades the market has dictated how we are to live. Through self-sufficiency we are able to design a life on our own terms. It is time to rethink the way we live – to move away from materialism and consumption as if they are the only way to a life worth living.

Self-sufficient homemaking is not just about chores and husbandry, it is also about the more romantic notion of homemaking – making a place of security and freedom. When you do it yourself you get to experience the satisfaction of independence by working for yourself and your family. Your days may be filled with gardening, animal husbandry and firewood collection. Or they might be taken up with the 101 other things that you could do.

Perhaps you have less time to devote to a self-reliant life, but you can spend your evenings and weekends pursuing the dream. So where do you start? Anywhere you can! More seriously, decide what is most important to you – fresh vegetables or organic meat – unique clothes, preserves or pickles. You will not only experience the satisfaction of doing it yourself but also get to enjoy higher quality goods.

Self-sufficiency offers us the chance to experience abundance in the form of fresh food, clean air and a healthy lifestyle. You will probably eat better than you ever have before. You will have fresh berries, asparagus and stone fruit in the summer. Apples, oranges and passionfruit in autumn. The point is that you will have a myriad of fresh fruit and vegetables in season if you grow them yourself. Summer is a very busy time in the self-sufficient household. It is time to prepare and preserve for the coming months so you can enjoy the fruits of your labours all year round. Pickles, chutney, sauces and bottled fruit, dried beans and frozen meat and vegetables all need to be prepared depending on your individual tastes. Perhaps you want to try your hand at winemaking from grapes or fruit. Perhaps you'll want to brew some beer. Do whatever suits you and your family.

Self-sufficient house-husbandry is a way of regaining old skills that have long been forgotten. By reskilling, and by doing it ourselves, we become more self-reliant and prepared for any emergency that might mean traditional product processes breakdown or become impractical.

You will never be bored or without something to do. In an attempt to remove oneself from the consumer market, there are a myriad of things we can do to gain independence and freedom. We might work less for money from conventional sources in order to spend our time in more profitable pursuits.

And that, for the self-provider, is the path to satisfaction and fulfilment. For the person caught up in materialism's superficiality and banality, self-sufficiency offers the proponent freedom and autonomy to choose a more meaningful lifestyle, to design a life worth living. This is one of benefits of self-sufficiency.

Self-sufficiency is all about designing a lifestyle that suits our purposes. But what else can do we do? We can cook, preserve, sow and grow. We can sew, knit and build, weave, raise animals, dig and paint. We can do so many things that bring meaning to our day, and the next day and the day after that.

The self-provider is not afraid to start from scratch. Buying an unworked piece of ground is an exciting process. And it could be your path to freedom. Preparing the soil with compost and manure to growing a vegetable garden to cooking up a stir-fry for dinner may be your ticket to liberation. Getting a few hens and a rooster, putting in fruit trees and perhaps some grapes might be just what you need to gain a sense of personal achievement.

If you can't keep animals in your high-rise apartment you could grow some vegies, herbs and maybe a lemon tree in a pot. Perhaps you might like to make your own clothes, preserve some fruit and try your hand at propagating your own seedlings. It is possible to be a bit self-sufficient anywhere. The more land you have the greater the opportunity to imagine a life that is more meaningful than a round of golf or a night watching reality television.

It is not my intention here to extoll the virtues of a

self-sufficient lifestyle while denigrating other choices. However, self-sufficiency can be more fulfilling than a life in the suburbs in which all chores are outsourced and costs an ever-increasing amount of money. It might suit you to 'insource' your gardening, house repairs, cleaning and making your own clothes, etc., etc.

There are many how-to manuals setting out how to live the self-sufficient life, raise animals and build your own dwelling (some of them are listed at the end of this essay). This essay is a 'why-to' rather than a 'how-to'. If you find routine mundane and outsourcing unfulfilling, self-sufficiency might be for you. If you get no satisfaction from your day job, if you long for autonomy in your everyday life, self-sufficiency might be the answer you are looking for. Self-sufficiency is not about doing it all. Rather it is about doing enough, and how much is enough is up to you.

In reality, a self-sufficient life is more complex than a lifestyle based on a series of consumer choices. It is, for now, a conscious decision, one that we have control over. We can choose to be self-sufficient. However, it appears that it is only a matter of time before it will become a necessity. As instability grows as a result of climate change, the more we will have to be reliant on our own resources.

It is likely that when we begin on our self-sufficient journey we might lack the skills to perform every task we want to. Part of the fun and sense of achievement comes from learning new skills and completing new tasks. As we gain confidence by gaining new skills we can try out new ideas. And we will increase our skillset so we are able to do more things and increase our self-sufficiency. The benefits are endless.

But we must go forth into our self-sufficient lives with our eyes open. The experience of living off and from the land (no matter how big that piece of land is) is anything but simple.

Hardworking advocates of self-sufficiency find that days are long. The potential outcome of a self-sufficient life is that the work is never done. But Megg Miller, as if speaking on behalf of self-providers everywhere, eloquently surmised in the 1970s: 'You don't mind doing these things – because you're in there, hands and feet, not outside it, not separated by layers of plastic, or automatic switched-on instant gratification'.*

Things seemed somewhat simpler in the 1970s. However, the drivers pushing people towards self-sufficiency remain the same today: the desire to be free from meaningless work, to be free of bosses that care more for their bottom line than the welfare of their workers, and alienation from the things that really matter (like healthy food, clean

---

* Meg Miller, in Margaret Smith and David J. Crossley, *The Way Out: Radical Alternatives in Australia*, Melbourne: Lansdowne, 1975, p. 197.

air and more time with family). Self-sufficiency still offers a solution to all these problems.

The philosophy of self-sufficiency has been criticized for being impossible for proponents to live up to. Few of us can drop everything and live like hermits in the country – even if we wanted to. But self-sufficiency does not require us to do so. Instead, it is about removing oneself from the consumer market as much as one can. That might mean living in the country or the city, or even living like the Nearings or the Seymours on the land. But it is equally possible to live self-sufficiently in the city. The more one can do the more self-sufficient you are – it is not an all or nothing lifestyle. You have not failed if you only grow some of your own veggies and do some craft to lessen your reliance on the consumer market and impact on the planet. If it takes you away even a little bit from the consumer market and materialism any gesture is welcome.

Self-sufficiency is not a simple lifestyle and, indeed, it is not promoted because it is. The further one takes it, the more one realises that it is hard. But it can be very satisfying.

A self-sufficient home could be anything from five acres and a cow to pots of fruit and vegetables on your balcony. There is always work to be done when you want to live self-sufficiently. There is always more you could do in an effort to be self-sufficient. As they say, from little things big things grow. Self-sufficiency is like that.

One might start out with a pot of tomatoes and end up on that farm with a cow. The possibilities are endless. You have to start somewhere – who knows where it will lead? Many of us who seek a self-sufficient life dream of five acres and a cow or a couple of goats. Or even one acre and a few chooks. It is a particularly empowering lifestyle for those of us who value independence and freedom.

But it is a lifestyle that has its caveats. It is often particularly difficult to say no to ever more work when it enhances your lifestyle or increases your self-sufficiency. There is the great tendency to acquire more animals, plant more crops, and build more in order to become more independent from the capitalist system. It is not unusual to find that your cow produces too much milk for your family. Your solution is to get a pig! But pigs need to eat things other than milk. And they need shelter. So you grow more food and build a sty. Maybe that pig has piglets and you need to feed them too. And what do you do with all those piglets? You can't keep them all. You can't eat them all. You might sell some in order to pay for basic necessities that you cannot produce yourself like fuel, medication and vet bills. The point is that work begets more work. If you have a milking cow there are times when the cow is dry and producing no milk. The solution is to buy your milk or to buy a goat (but a goat is a social animal so you may need to buy two). You choose goats; and so the cycle continues. And before long you have yourself a thriving menagerie, an even bigger vegetable garden and paddocks full of fodder crops.

I write this not to put off those seeking self-sufficiency but to highlight the necessary consequences of doing it all. Many proponents of self-sufficiency advise starting small. But often the desire and the necessity comes before the practicalities.

Not all can have five acres and a cow. Indeed, not since the 1970s has this been possible for the majority of those seeking self-sufficiency. There are many obstacles standing in the way of those who want to get away from it all to live a life in the country. Lack of or distance from employment is one; distance from friends and family is another. There are council regulations to contend with and neighbours to accommodate.

By doing as much as you can necessarily results in reducing the family budget. If you start small you can save yourself money. Rather than deciding to do it all straightaway do it bit by bit and build on that.

If you want to buy a whole lot of consumables, self-sufficiency is not for you. Many aspects of a consumer lifestyle are simply not compatible with self-sufficiency. And that, for many, is entirely the point.

There is nothing conventional about doing it yourself. Indeed, self-sufficiency offers a radical alternative to the mundane round of shopping, office work and all the other things that fit in with a materialistic life. You are free to do your own thing regardless of your economic circumstances, your geographical location or your personal means. Anyone can do it themselves, even it is just a little rather than it all. As I have argued, self-sufficiency is about doing what you can and where and when you can. It is not about doing it all (but you can try if you want to).

Self-sufficiency means, in essence, independence from traditional economic structures and self-reliance and autonomy. By being self-sufficient in any way you can, you can beat the marketers at their own game by producing what you really want, rather than buying what some third party wants to sell to you. If you want new clothes you can make them yourself. If you want fresh food you can grow it, process and cook it yourself. If you want to raise your own animals – you can do that too. In the self-sufficient household there are so many things we can do – what they are is up to you.

CONCLUSION

In this essay I have argued that self-sufficiency offers a solution to the 'problem of production' as raised by E.F. Schumacher. But it does much more than that. Self-sufficiency is about independence and freedom from the capitalist system. It is also a way to remove us from a materialistic lifestyle that has many negative consequences not least the destructive impact on the environment and the direct links to climate change.

Importantly, because self-sufficiency is about making oneself both the consumer and producer, marketers are less able to sell back to us the perceived benefits of the lifestyle. In essence, marketers cannot co-opt the central tenets of self-sufficiency to entice us to buy into a lifestyle that depends on removing oneself from the consumer market. Because self-sufficiency is at odds with capitalism, marketers simply do not know how to invade it and corrupt it.

Many people, of course, search for the simple life only to discover that it is anything but simple. Yet, there are always choices about how we are to live. Some choices have serious negative consequences and some are obviously better than others. And, for the planet some are clearly a no-brainer. Self-sufficiency is the ultimate act of rebellion, and one that could make a real difference to the *status quo*. So, let's simplify.

It is a strange kind of consolation that the economic prosperity witnessed over the last two hundred years, and particularly the last sixty years, has resulted in real and expected environmental catastrophe. Over-consumption is at once revered (by capitalists) and scorned (by the rest of us). As prosperity expanded in the years following the Second World War consumers were elevated in status. Consumers, acquiring rights never experienced before, wielded unbridled power. Yet this new power has never been fully utilised, as Australian consumers have been largely unwilling to mobilise as a group. Consumer boycotts are not a significant feature of the Australian consumer scene. Despite this, consumers remain a potential political force to be reckoned with. Self-sufficiency requires an individual response and one that is compatible with the model of the consumer that modern economics has produced. It is not, however, compatible with consumerism or materialism. Individuals can make a real difference.

The environmental catastrophe that is anthropogenic climate change has been the subject of significant concern for more than forty years. Yet despite growing interest, and almost universal acceptance of climate change by climate scientists, lack of real action has seen the issue stagnate for a number of reasons not least the prioritisation of the growth model of economics, the advance of globalisation, and burgeoning free trade agreements.*

Within this power structure individual action seems to be a waste of time. We need to step away from the consumerist system to achieve a revolution in values to truly reform the capitalist system. Despite our growing interest in 'green' solutions, our global ecological footprint is simply too high. Even if we significantly change our consumer behaviour and personally reduce CO2 emissions, by following Al Gore's advice and put more solar panels on our roofs, we are not doing enough. Even if we insulated our homes better, drove more efficient cars, and became vegan, we are not doing enough. Even if we change our behaviour as consumers we are not doing enough to combat climate change. Indeed 'enough' is not a word often uttered in the world of marketing and advertising. And, we cannot wait for Government action however we may need it.

As Richard Heinberg insightfully wrote: we have reached 'peak everything'.† Indeed, we should not have to rely on individuals voluntarily reducing their carbon footprint to

* Klein, *This Changes Everything.*

† Richard Heinberg, *Peak Everything: Waking Up to the Century of Declines*, Gabriola, BC: New Society Publishers, 2007.

deal with the enormity that is climate change and global warming. But, it seems, individual action is all we have.

Climate change is a peril of prosperity. Unbridled capitalism is responsible for untold harm not least harm to humans, animals, and the planet, harm that is likely to increase exponentially. The climate crisis is not an issue of future imaginings, it is here and it is real. We must radically change our consumer behaviour, our lifestyles, and our expectations about luxuries and necessities. Without doubt we need top-down (state) responses to the climate crisis. We need governments to reign-in consumer rights and marketers and advertisers to stop driving the economic agenda.

But we are still waiting.

In the aftermath of the Paris Climate Conference 2015 we still hold our breath for serious emission reduction solutions. When the Federal Treasurer Scott Morrison handed down his Budget 2017–18 on Tuesday 9 May 2017, he did not mention climate change, global warming or renewable energy. Not a single mention and not a single cent to be spent on climate change mitigation or adaptation. There were of course, cuts to be made to the funding of groups concerned with climate change. On 1 June 2017 President Donald Trump pulled the United States out of the Paris Climate Agreement. Despite its advancements in alternative energy technologies China remains the largest carbon emitter in the world. Australian Prime Minister Malcolm Turnbull reiterated his government's commitment to the Paris Agreement at the same time as he gave the Adani coal mine the go-ahead. Adani is set to establish Australia's biggest coal mine in the Galilee Basin in Queensland to extract coal for export to India. Given that the Queensland and Turnbull Governments have given the go-ahead to this project and to the expansion of the Abbot Point coal terminal suggests that their interests lie elsewhere.

While it is still unclear whether Adani will be able to find financial backing for the project, climate change is clearly not on the Australian or United States' agenda. It is not clear whether the coalition government is looking for private sector funding of geoengineering (for better or worse). It is not clear whether there are many votes in action on climate change. It seems not; at least not yet. But can we wait?

We will have to simplify regardless of political outcomes. Let us simplify sooner rather than later.

Let us simplify now.

FURTHER READING

CLIMATE CHANGE

John Broome, *Climate Matters: Ethics in a Warming World*, New York: Norton and Company, 2012.

Naomi Klein, *This Changes Everything: Capitalism vs. the Climate*, New York: Simon and Schuster, 2014.

Clive Hamilton, *Requiem for a Species: Why we Resist the Truth about Climate Change*, Crows Nest (NSW): Allen and Unwin, 2010.

David Suzuki and Ian Hanington, *Just Cool it: The Climate Crisis and What We Can Do*, Sydney: New South, 2017.

SELF-SUFFICIENCY AND SIMPLE LIVING

Samuel Alexander and Amanda McLeod (eds.), *Simple Living in History: Pioneers of the Deep Future*, Melbourne: Simplicity Institute, 2014.

Michael Allaby and Peter Bunyard, *The Politics of Self-Sufficiency*, Oxford: Oxford University Press, 1980.

Sharon Astyk, *Depletion and Abundance: Life on the New Home Front*, Gabriola Island (British Columbia): New Society Publishers, 2008.

Dona Brown, *Back to the Land: The Enduring Dream of Self-Sufficiency in Modern America*, Wisconsin: University of Wisconsin Press, 2011.

Linda Cockburn, *Living the Good Life: How One Family Changed Their World From Their Own Backyard*, Prahran (VIC): Hardie Grant, 2006.

Helen and Scott Nearing, *The Good Life: Helen and Scott Nearing's Sixty Years of Self-Sufficient Living*, New York: Schocken, 1970.

John Seymour, *Farming for Self-Sufficiency: Independence on a 5-acre Farm*, London: Faber and Faber, 1976.

John Seymour, *The Fat of the Land*, London: Faber and Faber, 1961.

Keith and Irene Smith, *The Hard Times Handbook*, Melbourne: Thomas Nelson, 1984.

John Vivian, *The Manual of Practical Homesteading*, Emmaus, Pa: Rodale, 1975.

CONSUMPTION

Clive Hamilton, *Growth Fetish*, Crows Nest (NSW): Allen and Unwin, 2003.

Amanda McLeod, *Abundance: Buying and Selling in Post-war Australia*, Melbourne: Australian Scholarly Publishing, 2007.

Vance Packard, *The Waste Makers*, New York: IG Publishing, 1960.

Guy Pearse, *Green Wash: Big Brands and Carbon Scams*, Melbourne: Black Inc., 2012.

David T. Schwartz, *Consuming Choices: Ethics in a Global Consumer Age*, Maryland: Rowman and Littlefield, 2010.

Printed in Australia
Ingram Content Group Australia Pty Ltd
AUHW020855090724
396769AU00003B/13